Ma'at of Data

AF820977

Cshantay Publishing
ISBN 978-1-7321319-8-9
Copyright 2024
All Rights Reserved

The Ma'at of Data

By

LaShonda C. Henderson MSc

I dedicate this book to all those who go unheard at the decision table. This is my rally cry to innovate. You are only bound by the words people say in the room. Out here in the real world, you can do it your way. I'm giving you this imperfect path, so you can perfect your own way.

We need your voice

Sound off

Loudly

Note from the Author

Since I was a child in grade school, I've been captivated by data and the stories it tells. As a young girl in rural Alabama, I found myself drawn to information and facts, eventually serving as Vice President of my high school's library club. The Dewey Decimal System wasn't just a method of organization to me—it was an art form.

My fascination extended beyond mere numbers. I wanted to understand the human experience through data: how many people were involved in historical events, what languages they spoke, and whether those languages were thriving or fading. This passion was fueled by my brother's enthusiasm for the Almanac and the Guinness Book of World Records. While he'd spout off city populations after watching a TV show, or hearing a conversation on the radio, I focused on organizing information to broaden my perspective beyond our small town.

Growing up in the late '70s and early '80s, we were witnesses to dawn of the computer age. I was thrilled to play Oregon Trail in computer class and ecstatic to take Typing 101, which allowed me to transfer my thoughts into a digital format quickly. When my aunt got a word processor, I pored over the manual, determined to understand how it operated without breaking it.

Though quiet by nature, my mind was always buzzing with questions and cross-checking facts. Our local and school library's limited resources often left me skeptical, teaching me early on that history and world events are often presented from an untruthful, unjust viewpoint. This realization would shape my approach to data management throughout my career.

My world expanded when I moved to a larger rural city and saved up for my first computer, from my job at the local sandwich shop. I'll never forget, it was a Gateway that came with a free disk that advertised 750 hours of free internet. In those dial-up days, if you wanted to be on the internet, you couldn't be on the telephone. So, I'd disconnect from the physical world to connect online, I found it to be a smorgasbord of information and wanted to keep it all! I learned quickly to save and archive fleeting web information using Lotus 1-2-3, an early software program. This was my first emotional dive into data management and information stewardship, organizing both structured and unstructured data to meet my insatiable appetite for learning. We are talking information from chat room scholars and the unlimited rare books available!

In college, I pursued a minor in Sociology and took a course in Archival Studies. My first real-world data management project was a graded assignment to organize my local church's office using records management principles. Armed with highlighters, labels, and filing cabinets, I created a system for tracking information—a skill I'd refine in subsequent roles at a tax preparation office and a mental health facility.

My combination of Southern hospitality, computer skills, and organizational prowess eventually led me to Washington D.C., where I started as an office secretary with the federal government. Over two decades later, I've been at the forefront of data management evolution in government agencies.

I've witnessed the establishment of Chief Data Offices across federal departments, driven by the Evidence-Based Policymaking Act of 2018. This legislation aims to improve the use of data in policymaking and program evaluation. I've had the privilege of designing data infrastructure for a federal headquarters agency and supporting two Chief Data Officers in advancing their data strategies. Throughout these experiences, I've championed methodical master data management with repeatable processes, always keeping the human element in focus.

What sets my approach apart is the combination of technical expertise with what I call "data common sense." In this book, I aim to show you how to understand data practically while maintaining a commitment to human-centered design. My perspective is shaped not only by years of experience but also by the principles of Ma'at, the ancient Egyptian concept of truth, balance, and justice.

By sharing my journey from a curious child in Alabama to a data management professional in Washington D.C., I hope to guide you in organizing and defining your data in a way that's both efficient and mindful of the people behind the numbers. Whether you're new to data management or a seasoned professional, this book will offer insights into creating data systems that are not just technically sound, but also aligned with human needs and experiences.

Join me as we explore how to harness the power of data while never losing sight of its ultimate purpose, **to improve our understanding of the world and each other.**

Introduction: Context

How to use this Book

Each section is this book will explore how to effectively engage with Data Management and Master Data Management (MDM) through a structured approach that blends storytelling, practical analysis, and reflective conversation.

We'll start by bringing data principles to life with a story, offering a vivid illustration of operations in action. Next, we'll break down the story's insights into actionable steps, guiding you on how to implement the principles in your own business. Finally, we'll prompt you to reflect deeply on your current data practices and envision the changes needed to align with these principles and provide some recommended KPIs, because every organization loves milestone checks!

Here is what it will look like in action:

1. Storytelling: Illustrating the Concept

Each section of this book will begin with a compelling story designed to bring the principles of data to life. This story serves as a metaphor for how data management and MDM work in a real organization.

2. Analytical Application: Translating Story into Practice

Following the story, each chapter will provide an in-depth analysis of how the principles illustrated, Ma'at and data management, can be applied to your organization. We will break down the story's lessons into actionable steps for implementation in your business. For instance, we'll explore how defining key data entities, establishing governance policies, and integrating systems can help you achieve a unified view of your data. You will learn how to tailor these principles to fit your organization's unique needs, ensuring that your data management practices support your overall business strategy.

3. Reflective Conversation: Deepening Your Understanding

To close each section, we will engage in a reflective conversation designed to prompt you to think more deeply about the actions you need to take. This will involve questions and exercises that encourage you to consider how the principles of data management apply to your specific context. For example, you might reflect on how your current data practices align with the principles of truth, justice, and balance, and identify areas for improvement. This introspective approach will help you crystallize your understanding and develop a clear plan for implementing a data infrastructure in a way that truly supports and enhances your organization's goals.

And finally, I will provide frameworks and resources to get you started on your own organizational journey.

By integrating storytelling, practical analysis, and reflective conversation, each section of this book guides you in an engaging and actionable manner, setting you on a path to a more organized, efficient, and harmonious data management system. Organized by the principles of Ma'at, grouping sections by three principles (truth, justice and balance). This approach encourages seeing the big picture. Let's journey!

Chapter 1: Awareness

Ma'at and Data what's the deal?

Organizations today face significant challenges in creating data governance structures that work, much less ones that scale with their mission needs. As they navigate an ever-evolving landscape of mandates, privacy concerns, and a multitude of new regulations and tools, it's easy to become entangled in technical minutiae. ***In doing so, we risk losing sight of the fundamental purpose of data: to make sound decisions that serve humanity or, depending on the organization's moral compass, potentially exploit it.***

Let's start with a simple truth: organizations, whether they're Fortune 500 companies or local non-profits, are essentially groups of people working together. They exchange services, information, and resources, all in an attempt to improve their lives and the lives of others. It's a dance as old as humanity itself, expanding from physical onto a digital stage.

Just as each individual is guided by a personal moral compass, an organization should be driven by a core ethos too. This guiding ethos shapes how it operates internally and interacts with the wider world. Aligning data design with a human-centered approach becomes essential in this context. By addressing the needs of the "whole" person, data transcends mere functionality and serves a higher purpose - it becomes a scorecard for humanity, communicating how programs and operations contribute meaningfully to the organization's mission and the well-being of the community served.

Aligning data with the principles of Ma'at represents more than a nod to proven wisdom; it's a passionate commitment to achieving true harmony and balance in our modern world. Rooted in the profound traditions of the ancients, Ma'at embodies the essence of truth, justice, and order, working together to synergize cosmic balance. By integrating these timeless principles into our data practices, we ensure that our information reflects accuracy and honesty while fostering fairness and equity.

Just as Ma'at maintained the delicate balance of the universe, aligning our data with these principles establishes a foundation of integrity and purpose. **Give people purpose, and their work becomes meaningful.** This approach guides us toward a more harmonious and just society, where every piece of information contributes to clarity, promotes justice, and sustains equilibrium in an ever-evolving world.

Here's what Ma'at encompasses:

Truth: At its core, Ma'at signifies unwavering truthfulness and integrity. It emphasizes that our actions and words must align with a universal truth, reflecting honesty in all aspects of life.

Justice: Ma'at embodies fairness and equity, ensuring that our actions are just and that we hold ourselves accountable. It represents the ideal of equitable treatment and the pursuit of justice for all.

Cosmic Order: This principle reflects the natural order of the cosmos, maintaining balance and harmony throughout the universe. Ma'at signifies that everything should exist and function according to its rightful place and purpose.

Ethical Conduct: In ancient Egyptian society, Ma'at guided moral behavior and judicial processes, ensuring social harmony and personal integrity. It was essential for upholding ethical standards and fostering a just society.

Here's a target state of Ma'at's usage in data management:

Truth: In data terms, this means ensuring our information is accurate and honest. No fudging the numbers or cherry-picking stats. When we commit to truth in our data, we build trust and make better decisions.

Justice: This principle pushes us to use data fairly and equitably. It's about asking tough questions: Are we collecting data ethically? Are our algorithms biased? Are we telling a data story that is based in truth and actionable?

Cosmic Order: In the data world, this translates to creating systems that work in harmony with human needs and natural processes. It's about designing data structures that reflect the real world, not forcing the world to fit our data models.

Ethical Conduct: This guides how we handle data, how we respond to data breeches. It's about respecting privacy and only collecting what is needed for the service provided, being transparent about data usage, and always considering the human impact of our data practices.

You may be thinking, what a load of crap, this sounds very fairytale-ish. Since you're already thinking in that manner, why not head full speed down that road! Have I got a story to tell you!

The Tale of Harmony Village

In a radiant land where the sun's golden rays kissed the hills and the rivers sang in harmonious tones, there was a village known as Harmony. This village was renowned for its serene beauty and the wisdom of its inhabitants. The secret to their peaceful existence lay in their unique approach to design, guided by the principles of Ma'at and the art of human-centered design.

The villagers, a diverse group spanning all ages, had long believed in the balance of truth, justice, and order. Their approach to creating spaces and solutions for their community reflected these values. One day, the village council decided it was time to build a new communal space—a place where everyone could gather, share stories, and celebrate together.

Empathy: To begin, the council knew they needed to understand the varied needs of their community. They held open forums where villagers from all walks of life, children, adults, and elders could voice their desires for the new space. Children wanted areas for imaginative play, adults sought spaces for meaningful conversations and work, and the middle-aged desired zones for relaxation and reflection. The council listened carefully, also observing how technology was integrated into their daily lives, from simple tools to advanced innovations that made their lives easier.

Iteration: With these insights, the council created initial designs for the communal space. They included elements such as interactive play zones with digital interfaces for children, multifunctional rooms equipped with the latest technology for adults to work and collaborate, and quiet, serene areas for relaxation for the middle-aged. The council built a small prototype, incorporating features like touch-sensitive tables, virtual reality corners, and smart lighting. Villagers tested these features, offering feedback that led to several refinements. Each iteration made the space more inclusive and functional.

Inclusivity: Ensuring accessibility was a priority. The space was designed to be welcoming to everyone. For the elderly or those with mobility challenges, there were ramps and adjustable seating. Children had interactive, safe play areas that were both engaging and educational. The integration of technology was mindful of all age groups—easy-to-use interfaces for the young, advanced tools for adults, and helpful, intuitive features for the middle-aged. The design aimed to bring all villagers together, respecting their diverse needs.

Collaboration: The project was a community effort. Artisans, engineers, and tech specialists worked alongside the villagers. Children's ideas for digital play spaces, adults' suggestions for work and meeting areas, and middle-aged villagers' input for relaxation zones were all integrated. Collaborative workshops allowed everyone to contribute, ensuring that the final design was a true reflection of the community's collective vision.

Problem-Solving: The council faced challenges, such as integrating cutting-edge technology while staying within budget and ensuring it was user-friendly. They addressed these issues by leveraging local resources and expertise, finding innovative solutions like modular tech that could be easily updated. Their creative problem-solving kept the project on track and within reach for everyone.

When the communal space was completed, it became more than just a structure—it was a living testament to the village's commitment to balance, harmony, and respect for all its members. The space was filled with interactive play zones, multifunctional rooms, and serene relaxation areas, all seamlessly integrated with technology that made life easier for everyone.

Lessons of Harmony Village

The villagers of Harmony found joy and connection in their new communal space, which honored their diverse needs and embraced the principles of Ma'at - truth, justice, and cosmic balance. It was a place where every person—child, adult, and elder—could find their place, their voice, and their peace. The village thrived, demonstrating that thoughtful design and a commitment to human-centered principles could create lasting harmony and joy.

In the heartwarming tale of Harmony Village, we witness a powerful lesson about the importance of unity and collaboration in achieving true progress. **The village's success wasn't just about the advanced technology or innovative designs but about the heartfelt commitment of its elders and the collective spirit of the entire community.** The elders, acting as visionary leaders, understood that real advancement could only be achieved by bringing everyone—children, adults, and elders—into the design. They held open forums, embraced diverse perspectives, and made sure that every voice contributed to the village's development. This inclusive approach not only fostered a sense of belonging but ensured that every need was met, creating a truly harmonious and functional environment.

This same spirit of unity and collaboration is crucial in implementing data governance and master data management (MDM). ***Picture data stewards as the modern-day elders of your organization***, guiding the transition from fragmented, siloed systems to a cohesive, enterprise-wide approach. Just as the villagers pooled their knowledge and efforts to create a balanced and thriving space, data stewards, service components (Risk, IT, Data, Information and Privacy offices) and leaders must work together to unify diverse data sources into a single, reliable framework.

Mastering data management is more than a technical solution; it's a transformative journey that requires the active involvement of every stakeholder. Just as the elders involved the whole village to achieve a shared vision, organizations must engage all departments and teams in the process. This collaborative effort ensures that data is not only accurate and accessible but also valuable and meaningful across the enterprise.

Building a strong data infrastructure with the same dedication seen in Harmony Village means breaking down silos and **fostering a culture where data is viewed as a collective asset and the people are seen as <u>the most valuable</u> strategic asset.** Then creating an efficient approach that reflects the needs and values of the entire organization. **The story of Harmony Village passionately illustrates that progress is driven by unity and shared vision.** In your organization, let the spirit of collaboration and inclusivity lead the way, and watch as your data transforms into a powerful force for harmony and success.

The story of Harmony Village also offers a powerful testament to how shared foundational principles can drive meaningful progress. The village's success—shaped by the elders' inclusive leadership and the collective effort of its people—beautifully embodies the essence of Ma'at: truth, justice, and cosmic balance. As we move forward in this book, we'll explore how these timeless principles can profoundly influence organizational behavior and outcomes.

In the following sections and chapters, we'll dissect approaching data through the lens of Ma'at, harnessing its core principles as our guiding framework. Just as the elders ensured that every decision aligned with the village's greater good, we will see how truth can underpin the accuracy and integrity of our data, how justice can promote fairness and equitable access, and how balance can harmonize data systems across the organization.

Join me in this exploration as we apply the principles of Ma'at to organizational data design shaped by the values of truth, justice, and balance. We'll begin with a foundational conversation around MDM; trust me, it's going to be one heck of a ride!

Chapter 2: Foundational Concepts

Activities Required for Data Management:

Master Data Management

If you researched or polled the top consulting industry leaders and thought partners around Master Data Management (MDM) you'd encounter varying perspectives for each organization. But you'd find that each recognizes the impact of effective MDM in achieving the agency mission.

Master Data Management (MDM) is a comprehensive method of managing an organization's critical data to ensure a single, consistent, and accurate view of business-critical information. MDM involves creating a centralized, unified repository for master data, which includes the most important data available or a design that is considerate of potential future data collection. It considers entities within an organization, such as customers, products, employees, suppliers, and locations (which can be structured as domains). Good MDM enables the creation of policies that truly reflect your organization's state, allowing you to establish and measure key performance indicators based on data science principles.

The goal of Master Data Management (MDM) is to ensure data consistency, accuracy, and accessibility across all business processes and systems.

While Industry may all approach MDM with different emphases, all share a common mission: to harmonize and optimize the use of the data collected to serve an organizational purpose. Whether through technology, governance, or integration, we're all on the hunt to create a single, authoritative source of truth that enhances decision-making and operational efficiency. All organizations SHOULD aim to utilize MDM to its fullest potential, ensuring that data not only serves its purpose but drives meaningful, impactful outcomes for their stakeholders, **MDM is your FOUNDATION!**

Implementing MDM

Before any organization creates a data strategy, or spend dollars on tools, it should first establish an MDM plan that is a fluid ***living policy*** that drives the organization towards its mission to serve or inform.

If you're looking to begin with a strong foundation, start by building a data infrastructure that aligns with the principles of Ma'at—truth, justice, and balance. This means, **knowing your data and where it lives, as well as its purpose to your mission.** It is important to note, by ensuring data is truthful, promotes justice, and fosters balance; it sends an internal message to the workforce. Because how you treat the data, really reveals how you treat your work environment. Don't you know that humans are behind the information? Acknowledging their contributions is crucial for both good business practices and a harmonious workplace.

What could this look like in practice? By creating a Master Data Management (MDM) system that communicates your commitment to truth, you ensure that issues and challenges are reported honestly and addressed effectively. Promoting data justice means that all stakeholders, whether consumers or producers, can expect fairness in their interactions with your data. At its core, MDM drives harmony across the enterprise, that's people and systems.

To get to that state, organizations MUST define MDM well, leaving no room for ambiguity about the core concepts. Then allow the interpretation of the terms into daily workloads, **with a mandate to document terminology and the design requirements of any software utilized as tool.** Don't underestimate the importance of documentation. It is critical to understanding your systems and data purpose. Once more, champion documentation in the development of any system or software and write it into your contracts if you use outside support.

Ensure that you host training and learning sessions that inform and communicate how your organization will embody the terms in your daily work and the design of any future initiatives.

Defined MDM concepts are your established organizational guardrails that will lead to a culture of learning, reference, and confidence in managing programs that that not only support clear and fair decision-making but that also maintain the balance and integrity necessary for a thriving organization. In the next section, we will delve into the activities and processes that create the complete picture of MDM, outlining how each component contributes to a unified and effective data management strategy.

MDM Activities Activated

As we consider the activities required in MDM, imagine "data governance" not as some stuffy rulebook, but as the vibrant heartbeat of your data strategy. Picture "data quality" as a daily workout routine, flexing and strengthening your information muscles. See "metadata management" as the art of storytelling, weaving rich narratives about your data's journey.

Folks, we've been doing it all wrong. We've been treating these terms like dusty old tomes on a library shelf when **we should've been using them as instruction manuals for an epic adventure!**

In the pages ahead, we're going to flip the script. We'll explore how these key terms come alive when we treat them as actions, not just abstract ideas.

Ready to see these definitions dance? Let's get this show on the road!

Data Governance is the personal stylist that:

- Establishes policies, procedures, and standards for managing data across the organization.
- Its activities involve defining data ownership, access rights, and providing a framework to support data stewardship, including setting standards around compliance with legal and regulatory requirements.

Data Quality Management is the detox tea that:

- Ensures that data is accurate, complete, consistent, and timely.

- Activities include data profiling, data cleansing, data validation, and ongoing monitoring of data quality.

Data Modeling and Architecture is the blueprint that:

- Involves designing the data structure and architecture to support MDM, including defining data entities, relationships, and data flows.
- Ensures that the data model is aligned with business needs and can accommodate future growth. It is the roadmap for where data lives in your organization.

Metadata Management is the mile marker that:

- Manages the data about data, providing context and meaning to the data.
- Involves creating and maintaining metadata standards, selecting repositories, including documenting data definitions, data lineage, and data usage for information discovery.

Data Integration the matchmaker that:

- Facilitates the seamless flow of data between different systems and applications for the organization.
- Involves data extraction, transformation, and loading (ETL) processes, it is the key to move your organization towards real-time data reporting. Here's a hint, rejection logs and audit trails are your friend!

Reference Data Management is your cheat sheet:

- Manages the standardization and consistency of reference data, such as codes and lists used across different systems (e.g., zip codes, industry classifications).
- Ensures that all systems within the organization use the same set of reference data, preventing discrepancies.

Master Data Consolidation is the matchmaker that:

- Involves identifying, matching, and merging duplicate data records to create a single, accurate "golden record" for each master data entity.
- Ensures that there is one consistent version of the truth across the organization. Indexing with intentionality can reduce your cloud costs!

Data Lineage and Auditing is the family tree that:

- Tracks the origin, movement, and transformation of data as it flows through the organization.
- Provides transparency and accountability, ensuring that data can be traced back to its source and verified. It is your roadmap to resourcing!

Data Security and Privacy is the bouncer that:

- Protects sensitive data from unauthorized access, breaches, and misuse.

- Involves implementing data encryption, access controls, and compliance with data privacy and cybersecurity regulations. This is your investment to avoid data breaches.

Data Stewardship is your bestie that:

- Assigns responsibility for managing data quality, consistency, and governance within the organization.
- Data stewards are teammates to leadership. They know what's happening daily. They make or break your data management policies and resolve data-related issues.

Data Lifecycle Management is your planner that:

- Manages the lifecycle of data from creation to archival and eventual disposal.
- Ensures that data is used properly in all stages. It helps identify which information should retained for the appropriate period.

Cloud Data Management is the VIP concierge that:

- Involves managing data in cloud environments, including data storage, backup, and disaster recovery.
- Ensures that cloud data is integrated with on-premises systems and that cloud data management practices align with overall MDM strategies.

Here's the deal:

I've laid out the MDM landscape for you, painting a picture of what it's all about. Now, I'm not expecting you to read this book cover to cover like it's the latest page-turner (though if you do, I won't complain!). Nope, this book is your personal buffet - take what you need, when you need it and debate me to find what works for you!

Now let's dive into the first principle of Ma'at and the data activities associated with it!

Chapter 3: Principle 1

Truth

±

Story for Truth

Can you remember a time when organizations were family run, and members of the communities in which they operated?

Imagine stepping way back in time, to an era when businesses had secretaries seated behind grand wooden desks. Intimidating yet reassuring once you realized the person at that desk had your best interests at heart. This was a time when the office was a hub of personal interaction and trust, where secretaries like Lucy held the organization's golden keys literally and figuratively. They were the gatekeepers, the first line of contact for every piece of information that entered the office. Picture Lucy, a meticulous and personable secretary in a bustling doctor's office, embodying the essence of master data management long before it had a name.

Lucy was more than just an assistant; she was the guardian of information. She devised the methods for collecting and managing data, welcoming each piece with a personal touch and ensuring it was handled with care and precision. Her desk was not just a barrier but a gateway to an organized world where information flowed seamlessly. Armed with a manila envelope and flawless penmanship, Lucy maintained order, ensuring that everything was in its rightful place and easily accessible. As the business grew, Lucy's methods became the standard, and she trained new employees, embedding her approach beyond the organization's policy.

In Lucy's office, data management was not just about procedures—it was a reflection of the principles of Ma'at: truth, justice, and balance. **Integrity was the foundation; you wouldn't lie to Lucy, and she wouldn't lie to you.** Her office was a place where truth and honor were not abstract ideals but everyday practices. **Trust was built on these steadfast principles, creating an environment where promises were kept, clarity was paramount, and every action was driven by a deep commitment to truth.** Working with Lucy's team meant partnering with an organization that honored its values and upheld its promises with unwavering integrity.

Lucy's role was pivotal—she was the backbone of the office, balancing the business with grace and efficiency. Her methodical approach ensured that the office ran like a well-oiled machine, making her the unsung hero of data management. Her methods were so efficient, that she often trained others on her methods. She created a community of likeminded, solution-oriented people. Let's think about how, Lucy and her impact harmonized the data management practices with truth.

Analytical Application of Truth

Principle of Truth: The principle of Truth emphasizes the accuracy, reliability, and transparency of data. In Master Data Management (MDM), this principle highlights the importance of maintaining data integrity through precise data collection, storage, and retrieval processes. Activities under this principle ***focus on building a foundation of trust in the data used across the organization.***

In our story, Lucy, the diligent secretary, embodies more than just the keeper of information; she is a guardian of truth, guided by principles. As we explore how Lucy's approach to data management aligns with Ma'at's principle of Truth, we will examine the core aspects of Master Data Management (MDM): Data Governance, Data Quality Management, Metadata Management, and Data Modeling and Architecture.

Data Governance:

Lucy's desk was more than just a piece of furniture; it was the hub of her meticulously governed information ecosystem. Each piece of data that passed through her hands was handled according to strict protocols she had established, ensuring accuracy and consistency. Lucy's desk and her filing cabinets were the heart of her carefully managed information system. **Every person and every piece of information they shared were handled with the strict protocols embedded in her written and unwritten policies.** Her approach was methodical, performing each task the same way every time.

Her tools were simple yet effective: a filing cabinet, manila envelopes, labels, and legible penmanship. In today's digital world, the filing cabinet represents digital storage systems (such as lakes, warehouses, and lakehouses), the manila envelope stands for file types, labels correspond to digital metadata tags, and the system timestamp serves as penmanship. Lucy had rules (data governance) for them all, and your organization should, too.

Lucy's meticulous data governance mirrored the principles of Ma'at, where truth and integrity were foundational. In today's context, data governance involves setting rules and policies on digital platforms to manage data responsibly, fostering trust and transparency. Just as Lucy upheld truth and integrity with her physical files, these values can thrive in today's digital practices. Whether your organization is fully digital or transitioning to digital you NEED governance controls to ensure processes survive employee transitions and enterprise technology enhancements.

It is important to note that Lucy's hands-on, front-line approach meant she was deeply familiar with daily processes. This understanding allowed her to craft effective governance policies.

Many organizations struggle with data governance because they are disconnected from day-to-day operations and miss the small details impacting frontline workers. Consequently, governance policies can become detached from real-time work. *(As we mentioned in the opening of the book, get to know your data stewards, they are the front-line keepers of the data lifecycle and your organizational success.)* Operating from a foundation of truth, Lucy never lied, and neither should you. Acknowledge what you don't know, invest time and resources to engage directly with processes, perform case studies to document workflows, and develop robust, scalable governance. It may require an initial investment of time, money, or resources, but it will ultimately save your organization in the long run.

Lucy or your organization can write all the governance it wants, but if it is based on incomplete, junk data, then no matter how transparent or truthful the governance policy, it won't help your data to serve the need of your organization. As we delve deeper into the essentials of effective data management, it becomes clear that data quality is a critical component that ensures the integrity of the entire system.

Data Quality Management:

Lucy's attention to detail was legendary. She didn't just process data; she knew each stakeholder submitting the information. Her precise penmanship and meticulous organization set the highest standards of data quality from the start. Each piece of data felt valued and understood, as Lucy embodied the first and only line of defense for data quality.

In Lucy's office, the dimensions of data quality were her guiding principles. Accuracy was crucial—every piece of information had to reflect real-time interactions, ensuring nothing was misrepresented. Completeness mattered deeply; missing details could lead to errors, such as a misdiagnosis or an overlooked appointment slot. Consistency was vital; Lucy ensured every file followed the same protocols, keeping data aligned.

Timeliness was another key aspect; Lucy updated data promptly to reflect the current situation. Validity was essential; she checked that data met defined formats and constraints. Reliability was crucial—Lucy was trusted to maintain high standards, ensuring data credibility. Relevance was important; she included only the necessary details for the office's needs, avoiding irrelevant information like a patient's favorite color.

Uniqueness was maintained by preventing record duplication, while integrity ensured accurate representation of data relationships. Conformity was key—Lucy's adherence to established standards ensured seamless integration into the larger system. Through these dimensions, Lucy upheld a data quality standard that was the bedrock of her office's efficient and trustworthy operation.

In contrast, the Office of Management and Budget (OMB) form process, mandated by the Paperwork Reduction Act (PRA) of 1995, along with standardized digital intake forms used by state and local organizations, focuses on uniformity and efficiency over personal interaction. These forms streamline data collection across organizations, ensuring consistency and regulatory compliance.

While these forms rely on self-reported information, analyzing how this information is input can reveal areas for improvement. Observation is your money maker! By examining how predefined fields and checkboxes standardize responses, organizations can enhance governance controls and reduce errors in the self-reporting process. This approach minimizes subjective interpretation and boosts data consistency.

Standardized digital intake forms, while efficient, often lack the personal touch and relational aspect of data management that Lucy provided. They use algorithms and automated systems to process data quickly but may miss the individual context behind the information.

Both approaches aim to ensure data accuracy and integrity, but Lucy's method stood out for its personal touch and meticulous attention to detail. This commitment to quality reflects Ma'at's principle of Truth, where the accuracy of information is crucial. In today's world, Data Quality Management involves continuously improving data accuracy, reliability, and relevance, ensuring that every piece of information effectively serves its intended purpose. Lucy collected only what was needed to fulfill the mission, she knew where it was and what it was, and her team shared responsibility for maintaining data standards.

Metadata Management:

Metadata is data about the data to locate the data, a mouthful, right? Lucy understood that managing data wasn't just about knowing what information she had, but also about understanding its origins and how it would be used. Her system for organizing files was meticulous; she used a numbering and alphabetizing method to ensure every piece of data was easily accessible and correctly stored. If Lucy were working with modern cloud systems, she would apply the same principles to metadata management, capturing the name, date, zip code, date the file was created, the last update and the associated coding system for storage and retrieval.

While Lucy created her standard because it made file retrieval easy, she demonstrated that secretaries could be seen as the pioneers of business MDM practices. However, currently, there are working groups, and organizations that have established methods for standardizing metadata handling with frameworks and templates to glean from. As you engage these spaces remember to embody the principle of truth to find that process that fits your needs.

A great resource is the National Institute of Standards and Technology (NIST). NIST has defined metadata as "Information describing the characteristics of data. This may include, for example, structural metadata describing data structures (i.e., data format, syntax, semantics) and descriptive metadata describing data contents." For digital or physical files, metadata helps in data retrieval through classification and monitoring the movement of data in your organization.

In today's digital age, metadata management involves capturing and maintaining comprehensive records about data sources, usage, and lineage. Just as Lucy used tags and a coding system to organize physical files, modern metadata management includes details such as file names, creation dates, and storage locations. This approach ensures that every piece of data is documented and understood in its full context.

Aligning with Ma'at's principle of Truth, modern metadata management emphasizes clarity and transparency. The Open Data Act, which mandates that data be made more openly available to the public, embodies these values by promoting transparency and accountability. By maintaining user friendly metadata standards, organizations can ensure that data is not only accurate but also properly documented, reflecting its true value and origin. This approach fosters trust and integrity in data management, aligning with the principles of truth and transparency just as Lucy's method did in her time.

Data Modeling and Architecture:

Lucy's organizational savvy extended to her intentional approach in arranging her workspace, creating a logical and efficient system for handling information—what you might call data modeling or architecture. According to IBM, "Data modeling is the process of creating a visual representation of an information system to communicate connections between data points and structures" (2021). Meanwhile, Dataversity explains that "Data Architecture happens within a systematic framework, providing a foundation for people and systems to work with data" (2023).

Lucy's focus was not on technical specifications but on ensuring that data was stored accessibly and effectively, supporting the smooth operation of her office. Her "data modeling" involved carefully arranged filing cabinets and a clear system for organizing and retrieving documents. She labeled each file, assigned a numerical categorization method to identify file location, and sorted them alphabetically, ensuring that everything was in its right place.

In data modeling, the first step is to define what information needs to be captured and how it will be used. You start by creating a concept model to visualize how different data points relate to one another. Next, you build a logical model that outlines how the data will be organized, using tables and showing their relationships, without worrying about specific technical details. Finally, you develop a physical model that translates these ideas into actual designs and storage solutions compatible with your technology. When addressing technical aspects, choose the right tools for your need, design databases with your data stewards, optimize performance, select storage options, ensure security, and test and integrate the system to ensure efficiency.

Lucy's method can translate as groundwork for modern data architecture, including AI modeling. AI and machine learning rely on well-organized data to train algorithms and make accurate predictions. Lucy's practices, including consistent labeling, clear organization, and reliable retrieval, reflect the requirements for effective AI data preparation. Just as her system ensured data was accessible and meaningful, today's AI models depend on high-quality data to deliver insights and automation.

In essence, Lucy's straightforward approach to data modeling and architecture, placing things where they needed to be and labeling them for easy retrieval, supports both operational efficiency and advanced technologies like AI. Knowing what data you have and where it resides is essential for taking your data usage to the next level.

By focusing on Lucy's usage of data governance, data quality management, metadata management, and modeling/architecture, we find the foundation of truth as necessary to ensure data reliability and operational efficiency. In no step did Lucy have to compromise her integrity, nor tell anything less than the truth to ensure that the organization functioned well. You don't either!

Now let's dive into a reflective conversation to engage your organizational foundation in actions and truth.

Reflective Conversation

Is your organization operating from truth? It is so important that you perform a self-evaluation and if possible, bring in a consultant to gauge your organizational integrity. It could simply mean, assessing if employees feel safe when bringing challenges to leadership, knowing they will be resolved. Are front line workers treated like trusted advisors in the design of your processes? Take a good hard look at the cultivated environment and then publicly (in house) announce the results and the remediation plan. But most importantly build your truth foundation utilizing the following actions and metrics:

MDM Activity needed	Why it fits the principle of Truth	Actions for MDM Foundation
Data Governance	Establishes policies and standards for managing data truthfully.	Develop and implement documented data governance frameworks.
Data Quality Management	Ensures data is created and remains accurate, complete, and reliable.	Implement data quality tools and processes for continuous improvement.
Metadata Management	Provides context and understanding, to inspect and review data on demand.	Create and maintain a metadata repository that details data origins and uses.
Data Modeling and Architecture	Structures data in a way that supports the truthful representation of information.	Design data models that accurately reflect business entities and relationships.

Reflective Questions and KPIs:

1. **Data Governance:** How can you develop and implement a robust data governance framework to ensure accuracy and uphold truth in your data management practices? KPI recommendations to test accountability:

- Percentage of data governance policies implemented
- Frequency of policy reviews
- Number of staff trained in data governance

2. **Data Quality Management:** What tools and processes can you put in place to continuously monitor and improve data quality, ensuring that data remains accurate, complete, and reliable? KPI recommendations to test accountability:
    - Percentage of data errors identified and resolved
    - Frequency of data quality audits
    - Percentage of data records meeting established quality standards
3. **Metadata Management:** How will you create and maintain a comprehensive metadata repository to provide clarity and transparency about the data's origins and uses? KPI recommendations to test accountability:
    - Percentage of data assets with complete metadata
    - Frequency of metadata updates
    - Number of metadata-related issues resolved

4. **Data Modeling and Architecture:** What steps can you take to design data models and architecture that accurately reflect your business entities and relationships, supporting truthful and effective data representation?
    - Percentage of data models aligned with business needs

- Frequency of model reviews and updates
- Number of data integration issues identified and resolved

Implementation Tips

- **Integrate Regular Reviews**: Schedule periodic reviews of your data governance, quality, metadata, and architecture practices to ensure they align with evolving business needs and data management standards.
- **Promote Training and Awareness**: Ensure that all team members are aware of and understand the importance of data quality and truth. Regular training sessions can help reinforce these principles.
- **Encourage Transparency**: Foster an environment where data management practices are transparent and where staff are encouraged to report issues and suggest improvements.

By taking these actions and monitoring the suggested KPIs, you'll build a culture of truth that underpins your organization's data management practices, ensuring that data remains accurate, reliable, and aligned with the Ma'at principle of truth. In the next section, we'll dive into our next principle of Ma'at, Justice.

Chapter 4: Principle 2

Justice

∞

Spoiler alert: This section is a whirlwind of moving parts and technical wizardry (don't even think about calling it boring!). It's crucial though, so grab your coffee, buckle up, and dive in. We need data to be just, to help protect the people and get resources where they are needed. So, pay attention to these details! Trust me, it's worth every ounce of your attention!

Story for Justice

Do you remember when we used to open the newspaper and see headlines with numbers and facts, and because it was in bold print, we believed it? There was a time when information outlets had to validate their numbers and perform thorough research before stating a fact. We might have called them experts of the information. Numbers have always been powerful. They determine the food we buy, assess the players we watch in sports, and influence countless decisions in our lives. Data can be a catalyst for good or an antagonist for harm.

Think back to when numbers were used to paint a dismal picture of a community or space. The numbers said it, so that must be true. Right? Or wrong? Stewardship of data matters, because how can numbers be seen as valuable if no one validated them against the truth? Context can significantly impact how data is received and acted upon.

Carl and Janet, two dedicated professionals trained by the Legendary Lucy, had always been influenced by her principles of data management. Lucy's legacy was more than just a method; it was a commitment to truth and integrity. Inspired by her, Carl and Janet embarked on a mission to ensure that data was not only accurate but also used to promote justice.

Carl worked for a local government agency, while Janet was part of a nonprofit organization. Both were aware of the power of data and how it could sway public opinion and decision-making. Their roles required them to handle sensitive information and make decisions that could affect many lives. They knew that with great power came great responsibility.

One day, Carl was faced with a troubling dataset about housing conditions in a low-income neighborhood. The numbers suggested a worsening situation, but something didn't sit right. He remembered Lucy's lessons about context and integrity. He decided to dig deeper, verifying the data sources and considering the broader context. What he discovered was a discrepancy in how the data was reported. The numbers were accurate but had been presented in a way that exaggerated the severity of the situation.

Carl took his findings to the community leaders and proposed a revised analysis. His transparent approach and commitment to presenting the data honestly led to deeper understanding of the housing issue. The revised data helped the community secure appropriate resources and support, ultimately leading to a fairer distribution of aid.

Janet, on the other hand, was working on a campaign to promote equitable access to education. She was tasked with analyzing data on school performance across various districts. Janet was aware that data could be used to either highlight disparities or mask them. She applied Lucy's principles of

Truth by ensuring that the data she used was comprehensive and represented the true picture.

She found that some reports were skewed by missing data from certain schools, which could mislead stakeholders about the actual state of education in those areas. Janet advocated for the inclusion of this missing data and worked with her team to ensure that the final report was both accurate and fair. Her efforts led to increased funding and support for underrepresented schools, promoting justice in educational opportunities.

Carl and Janet's experiences highlighted how the principles of Ma'at—truth, justice, and balance—can be fully embodied in data management. Their dedication to validating data, understanding context, and using information ethically reflected their deep respect for the principle of justice. They showed that when data is managed responsibly and with integrity, it can become a powerful tool for driving positive change.

Their story underscores the importance of stewarding data with care and aligning organizational missions with broader community needs. As Carl demonstrated, the true value of data goes beyond mere numbers; it lies in how we leverage these numbers to uphold justice and foster equitable outcomes for all.

In the next section, we will explore our findings through a case study observation format. This approach will illustrate how organizations can examine their processes through the lens of justice, ensuring that their data management practices reflect these core principles.

Analytical Application of Justice

Data Lineage and Auditing Observation

Carl and Janet, guided by Lucy's expertise, approached their roles with precision. They knew that tracking data lineage was vital for maintaining transparency and accountability. By documenting each data point's journey from origin to final use, Carl avoided technical debt and ensured that data was used effectively throughout its lifecycle.

Importance of Tracking Data Lineage: Understanding how data moves allow for accurate assessment of data's age and produces optimal reporting. For organizations, having outdated data can be as problematic as having no data at all.

Auditing Practices: Janet's creation of detailed audit trails. Logging updates, deletions, and modifications enabled the organization to trace issues back to their source. Effective auditing saves on costly consulting fees and prevents work stoppages.

Implementation Steps:

Documentation: Document where data originates, how it moves, and where it is stored. This transparency helps in managing cloud costs and data protection.

Auditing: Conduct regular audits to ensure data accuracy and integrity, providing a clear understanding of data flows and transformations.

By maintaining comprehensive records and thorough auditing, organizations uphold the Ma'at principle of Justice, promoting transparency and accountability in data management.

Data Stewardship Observation

Carl and Janet embodied the spirit of data stewardship, which is crucial for ethical management. Carl focused on data quality and usage, while Janet developed and enforced data management policies.

Value of People: While data is a strategic asset, **the people managing it are even more valuable**. Building strong relationships with program managers and data stewards is essential. Rather than dictating and micromanaging, organizations should seek their insights and reward their contributions. It's important to recognize the consistent work of data stewards, not just high-visibility projects.

Effective Stewardship: Carl and Janet's commitment involved monitoring data quality, addressing issues, and providing training, fostering a culture of continuous learning then empowering responsibility and adherence to best practices. This is critical in an organization that has a mission that forever changes policy and is super prone to change.

Implementation Steps:

Empower Stewards: Assign dedicated data stewards, involve them in decision-making, and recognize their efforts. Cross train stewards. They are the key engines to your organizations, let the stewards know your commitment by assigning offices to act as a service in maintaining their operational processes. Make them feel supported and be open to their feedback as an advisory to all policy decisions.

Maintain Integrity: Ensure data handling aligns with guidelines and policies, promoting ethical management and high data quality.

Carl and Janet's approach to data stewardship reflects the Ma'at principle of Justice by upholding ethical management and ensuring compliance with policies. Their work in fostering responsibility and integrity is crucial for maintaining trust and effective data use. With the foundational aspects of data lineage, auditing, and stewardship firmly established, Carl and Janet turned their attention to another critical dimension of Justice in MDM: data privacy and security.

Data Privacy and Security:

Ensuring that data is protected against unauthorized access and breaches is essential to upholding the principle of Justice and maintaining the trust of both the organization and its stakeholders. Your stakeholders share information with you for an intended purpose and they expect you to utilize it for that only. It is unethical for organizations to sell information for profit or share it to increase sells or profit share. Customers are more likely to spend money with organizations they trust and operate with a sense of justice and fairness when it comes to protecting their information.

Carl and Janet were vigilant about protecting sensitive information, fully understanding the importance of data privacy and security. Carl implemented robust access controls to ensure that only authorized personnel could view or modify sensitive data. To mimic Carl's outcome and perform this work effectively, your organization MUST tag and categorize data into domains and categories. A solid taxonomy plan streamlines the implementation of access controls. With zero trust initiatives becoming more popular, the "trust no one and verify everyone" approach necessitates a structured methodology for grouping information, similar to the system Carl put in place.

Janet focused on encryption techniques to protect data during storage and transmission, while keeping the team informed about compliance with privacy regulations. She also developed procedures for managing data breaches, ensuring a swift and effective response to any incidents.

While many organizations concentrate on assembling the right technical team to reduce risks and develop software for privacy and security, **a crucial element is often missing: a policy analyst.** This role is vital for minimizing the impact of hacking and external threats. Good security and privacy measures require robust checks for vulnerabilities, which can only be achieved by pairing a policy expert with developers and risk management teams, dedicated to ensuring compliance.

Privacy and Security Measures: Carl and Janet understood that protecting data requires stringent security protocols and adherence to data protection regulations. They implemented comprehensive measures to safeguard data from potential breaches and ensure privacy compliance.

Implementation Steps:

Security Protocols: Enforce robust security measures to prevent unauthorized access and potential breaches.

Compliance: Ensure adherence to relevant data protection regulations and standards to maintain privacy and security.

By integrating these practices with their work on data lineage, auditing, and stewardship, Carl and Janet upheld the Ma'at principle of Justice, creating a secure and ethical data management environment. Their commitment to data privacy and security, further reinforced the integrity and fairness of the organization's data practices. Regular security assessments and vulnerability scans were integral to their routine, helping to identify and mitigate potential risks. For organizations, performing these activities is essential for protecting data from unauthorized access and breaches, maintaining customer trust, avoiding legal penalties, and ensuring compliance with data protection regulations. Through their diligent efforts, Carl and Janet exemplified the Ma'at principle of Justice, ensuring that data privacy and security were upheld to the highest standards.

Data Archiving and Retention

Carl and Janet approached data archiving and retention with the same meticulousness that characterized their work. Carl established clear policies for data retention, defining how long different types of data should be kept and when they should be archived or disposed of. Janet focused on secure archiving practices, ensuring that historical records were protected from unauthorized access while complying with legal requirements. They conducted periodic reviews to ensure that their archiving strategies remained aligned with evolving needs and regulations. When data was no longer needed, Carl and Janet followed secure disposal methods to ensure complete and irreversible deletion.

For organizations, effective data archiving and retention practices are vital for compliance with legal requirements, optimizing data storage costs, and ensuring that valuable historical information is preserved securely. Their commitment to ethical data management throughout the data lifecycle reflected the Ma'at principle of Justice, ensuring that data was managed responsibly and in accordance with legal and privacy standards.

Reflective Conversation

What does justice look like at your organization? Can your stakeholders trust you? Do your employees trust your organization's decision making. Is there open transparent communication happening? Are whistleblowers protected?

Justice: Activities should focus on fairness, accountability, and the ethical handling of data. These actions are vital for ensuring that your data management practices are responsible and compliant from the start.

MDM Principle	Why It Fits the Principle of Justice	Actions for MDM Foundation
Data Lineage and Auditing	Tracks data's origins and transformations, ensuring accountability.	Implement systems to document and audit data lineage for transparency.
Data Stewardship	Promotes ethical management and use of data.	Assign data stewards to oversee data handling practices and policy compliance.
Data Privacy & Security	Protects data against unauthorized	Enforce security protocols and comply

	access and breaches.	with data protection regulations.
Data Archiving & Retention	Manages data lifecycle ethically, respecting privacy and legal boundaries.	Establish policies for data retention and secure archiving in compliance with laws.

Data Lineage and Auditing

Reflective Questions:

1. How can we ensure comprehensive documentation of our data's origin and transformation processes? KPI recommendations to test accountability:

- ✓ Percentage of data sources with complete lineage documentation.
- ✓ Number of data transformations logged and reviewed.
- ✓ Frequency of updates to data lineage documentation.

2. Are our current auditing practices effective in identifying and resolving data discrepancies? KPI recommendations to test accountability:

- ✓ Number of discrepancies identified through audits.
- ✓ Average time to resolve data discrepancies.
- ✓ Percentage of audit findings that lead to actionable improvements.

3. How regularly do we perform audits to maintain data accuracy and integrity? KPI recommendations to test accountability:

- ✓ Frequency of scheduled internal and external audits.
- ✓ Percentage of audit schedules adhered to.
- ✓ Number of audit-related issues that result in policy changes or improvements.

Data Stewardship

Reflective Questions:

1. How do we ensure our data stewards have the necessary authority and support to effectively manage data? KPI recommendations to test accountability:
   - ✓ Number of data stewards with decision-making authority.
   - ✓ Frequency of support and feedback sessions with data stewards.
   - ✓ Percentage of data steward recommendations implemented.
2. Are our data stewardship policies and practices promoting ethical data management? KPI recommendations to test accountability:
   - ✓ Number of data stewardship policies reviewed and updated annually.
   - ✓ Employee feedback on the effectiveness of stewardship practices.
   - ✓ Compliance rate with established data management policies.
3. What methods are in place to recognize and reward the contributions of data stewards? KPI recommendations to test accountability:
   - ✓ Number of recognition and reward programs for data stewards.

- ✓ Frequency of performance reviews and acknowledgment of contributions.
- ✓ Employee satisfaction and retention rates among data stewards.

Data Privacy & Security

Reflective Questions:

1. How effective are our access controls in preventing unauthorized data access? KPI recommendations to test accountability:
    - ◊ Number of access control breaches reported.
    - ◊ Percentage of data categorized and tagged for access control.
    - ◊ Frequency of access control policy reviews and updates.
2. Are our encryption practices sufficient to protect data during storage and transmission? KPI recommendations to test accountability:
    - ◊ Percentage of data encrypted both in transit and at rest.
    - ◊ Number of incidents involving unencrypted data.
    - ◊ Compliance rate with encryption standards and regulations.
3. Do we have a comprehensive strategy for managing and responding to data breaches? KPI recommendations to test accountability:
    - ◊ Time taken to respond to and resolve data breaches.
    - ◊ Number of breach response drills conducted annually.
    - ◊ Percentage of breaches with documented response plans.

Data Archiving & Retention

Reflective Question:

How effectively are we managing the lifecycle of our data to ensure compliance, optimize storage costs, and securely preserve valuable information? KPI recommendations to test accountability:

- Compliance Rate: Percentage of archived data that adheres to legal and regulatory retention requirements.
- Cost Efficiency: Percentage reduction in storage costs due to optimized data archiving and retention strategies.
- Preservation Effectiveness: Number of instances where data was successfully retrieved and used from archives without data loss or corruption.

Implementation Tips

In today's data-driven landscape, ensuring the integrity, security, and effective management of data is paramount. To achieve this, organizations must implement rigorous practices across several key areas: data lineage and auditing, data stewardship, data privacy and security, and data archiving and retention. Each of these areas plays a crucial role in maintaining the balance between operational efficiency and the ethical handling of data. By focusing on comprehensive documentation, effective stewardship, robust security measures, and optimized data lifecycle management, organizations can uphold high standards of data governance.

Here are three high-level implementation tips to enhance these practices:

Develop a Comprehensive Documentation Strategy

- **Action:** Create and maintain detailed records of data lineage, including origins, transformations, and storage.
- **Why It Matters:** Ensures transparency and accountability, making it easier to trace data discrepancies and uphold data quality.
- **Tip:** Implement automated tools for documentation and regularly review and update data lineage records to capture changes accurately.

Enhance Data Stewardship Practices

- **Action:** Empower data stewards with clear authority and support and establish regular feedback mechanisms.
- **Why It Matters:** Promotes ethical management and efficient handling of data, ensuring that data stewardship aligns with organizational policies.
- **Tip:** Develop and enforce policies that recognize and reward data stewards and provide ongoing training to support their decision-making.

Strengthen Data Privacy and Security Measures

- **Action:** Implement stringent access controls and encryption practices and establish a comprehensive data breach response strategy.
- **Why It Matters:** Protects sensitive information from unauthorized access and breaches, ensuring

compliance with regulations and maintaining stakeholder trust.

- **Tip:** Conduct regular security audits and breach response drills to identify vulnerabilities and refine security measures continuously.

The implementation tips, grounded in the principles of data management, provide a solid foundation for organizations seeking to enhance their data governance practices and achieve a balanced, ethical approach to data management.

Just as Lucy embodied the principle of truth, Carl and Janet enhanced their core ethos by incorporating the principle of justice, further strengthening the foundation of truth.

Their application of Ma'at's sense of justice allowed them to use data to uphold truth, **champion the needs stakeholders to have unbiased data,** and respected the value of data stewards.

They ensured data was protected from misuse and maintained its integrity by knowing what was where and auditing appropriately. Carl and Janet's example illustrates how applying the principle of justice in data management upholds the integrity of information and the organization, ensuring it is both accurate and ethically managed.

Their story is a compelling example for organizations to integrate these justice-driven practices into their data management strategies, reinforcing the Ma'at principle of justice in their operations.

In the next section, we'll dive into our final principle of Ma'at, balance.

Chapter 5: Principle 3

Balance

Δ

Once things are working, it can be challenging to maintain a steady flow of processes that sustain that success. Organizations aim to harmonize data, but we should also focus on harmonizing the environment. Let's explore how to achieve that:

Story for Balance

In the heart of Harmony Village, where wisdom flowed as freely as the rivers and the knowledge of the elders guided every decision, Lucy, Carl, and Janet faced a new challenge. Their mission was to extend the village's exemplary data practices to new realms, ensuring that the principles of Ma'at—justice, truth, and balance—were deeply embedded in data management practices across the land.

The Call to Action

The village elders, impressed by the seamless integration and ethical data management within Harmony Village, saw an opportunity to share their success with other communities. They summoned Lucy, Carl, and Janet to develop a strategy for transferring these proven practices into broader contexts.

"Lucy," said Elder Amara, "you have upheld the principle of truth in our data practices. We need you to document how our practices can serve as a model for other communities."

"Carl," continued Elder Amara, "your technical expertise has been instrumental in integrating and managing our data. Your role is to create frameworks that illustrate how these principles can be applied in various settings."

"Janet," added Elder Amara, "your work in safeguarding data has been exemplary. You will be responsible for gathering evidence of the effectiveness of these practices and demonstrating their value to other organizations."

Crafting the Framework

Together, Lucy, Carl, and Janet embarked on their mission, guided by the wisdom of the village elders. Their task was to document and demonstrate the effectiveness of Harmony Village's practices in four key areas: reference data management, master data management consolidation, data integration, and cloud integration.

- **Reference Data Management:** Lucy began by detailing how a unified reference data system had improved data consistency and decision-making within the village. Feedback from villagers indicated that they could find what they needed quickly based on the legends and mapping created. She enhanced this process by documenting and identifying comprehensive guidelines that showed how maintaining a single source of truth could resolve data discrepancies and enhance accuracy in various contexts.
- **Master Data Management Consolidation:** Carl developed frameworks to illustrate the benefits of master data consolidation. He created a roadmap that was easy to read and follow, even the children understood what would happen when. He documented how consolidating disparate data sources into a single, accurate repository stream operations and improve reporting. His frameworks included practical

steps for other organizations that the village shared information with to achieve similar results.

- **Data Integration:** Janet focused on the seamless flow of information achieved through data integration. She provided evidence of how integrating data from multiple sources had eliminated silos, improved collaboration, and enhanced overall efficiency. Her reports included case studies and best practices for effective data integration to share with other organization. Creating industry information exchanges to promote promising practices.
- **Cloud Integration:** Carl and Janet worked together to showcase the advantages of cloud integration. They performed demos and created capability matrices to aid decision-making. These matrices included cost, risk, and change management levels of effort, along with capabilities. They documented how leveraging cloud solutions had enabled scalable data management, secure access, and enhanced collaboration. Their evidence demonstrated how cloud integration could support growing data needs and improve operational flexibility.

Gathering Evidence

The trio was on it! They meticulously gathered evidence to support their documented practices:

- **Reference Data Management:** Metrics and case studies showing improved data consistency and accuracy.

- **Master Data Management Consolidation:** Evidence of streamlined operations and enhanced reporting capabilities.
- **Data Integration:** Case studies and reports on improved information flow and collaboration.
- **Cloud Integration:** Metrics highlighting the benefits of scalability, security, and operational flexibility.

Sharing the Wisdom

With their findings and frameworks in hand, Lucy, Carl, and Janet presented their work to the village elders. The elders, recognizing the value of the documented practices, prepared to share these insights with other communities.

Elder Amara addressed the gathering, "The wisdom and evidence presented by Lucy, Carl, and Janet will guide us in spreading our successful practices. By documenting our promising practices, we ensure that our communities are solution-oriented and share what we have learned to empower others. The principles of Ma'at—justice, truth, and balance—are upheld in data management across the land."

Analytical Application of Balance

Guided by the wisdom of the village elders, in the tale of Harmony Village, Lucy, Carl, and Janet, thoughtfully applied principles of balance and justice to their data management practices. Their approach, where each member took responsibility and engaged stakeholders, offers valuable insights into how organizations can achieve balance and efficiency in managing their data across various domains.

Balance is one of the most challenging principles of Ma'at. Organizations must ensure they have what they need while positioning themselves towards being able to obtain future needs. All while having change management structures in place to reduce strain and burden on employees. This section differs from others, focusing on communicating key activities for organizations to take.

We'll start by defining the key activities needed, then provide analytical insights and recommendations for organizations to identify, maintain, and continue balanced actions in their environment. Here's an analytical breakdown of how these principles apply to reference data management, master data management consolidation, data integration, and cloud integration, along with actionable recommendations for organizations.

Ma'at seeks continuous balance and alignment. Create an environment where improvement matters.

Reference Data Management

Key Activities:

- **Definition and Classification**: Establishing clear definitions and classifications for reference data ensures consistency across systems.
- **Governance and Quality Management**: Implementing policies for governance and quality management maintains the accuracy and reliability of reference data.

- **Integration and Security**: Ensuring integration across systems and safeguarding against unauthorized access supports data consistency and security.

Analytical Insights: Lucy, Carl, and Janet's approach to reference data management emphasized the importance of defining and classifying data to achieve consistency. By setting up robust governance and quality management practices, they ensured that reference data remained accurate and reliable. Integration across different systems and strict security measures were vital for maintaining data consistency and protecting against misuse.

Recommendations for Organizations:

- **Define and Classify**: Clearly define and classify reference data to support consistency and cross use.
- **Implement Governance**: Establish governance policies and quality management practices to ensure you get the same outcome with each interaction with the data. Focus on data accuracy and reliability.
- **Integrate and Secure**: Focus on seamless data integration and security measures to protect data.

Master Data Management Consolidation

Key Activities:

- **Centralization of Data Sources**: Consolidating master data from siloed and program specific sources into a single, enterprise space for improved governance and auditing opportunities.

- **Data Quality and Consistency**: Ensuring high standards for data quality by setting controls and measure for consistency throughout the data lifecycle across programs and for a consolidated, trusted, quality master data source.
- **Governance and Stewardship**: Implementing governance frameworks and appointing data stewards to manage and oversee master data. Creating governance WITH stewards, and prescribing solutions together, ensuring quarterly communication channels to exchange feedback and offer change management after action reviews.

Analytical Insights: The consolidation of master data sources was crucial for achieving a single version of the truth. Using this method, organizations should strive to create an enterprise analytics platform where extraction, transforming, and loading exists in sandbox space to predict the quality and use of data. Canned reports and ad hoc reports should be recommended for reduced time spent in using big data to perform complex visualizations. By focusing on data quality and consistency, Lucy, Carl, and Janet demonstrated how consolidation can reduce discrepancies and enhance data reliability. Their governance frameworks and data stewardship ensured that master data was managed effectively and aligned with organizational standards.

Recommendations for Organizations:

- **Centralize Data**: Consolidate master data into a central repository to ensure a single source of truth. Centralizing data must include a documented path to communicate with data stewards to ensure that data usage stays in its proper context.

- **Ensure Quality**: Maintain high standards for data quality, design with quality and future state in mind. Communicate, Communicate, Communicate changes and consistency across the consolidated data.
- **Govern and Steward**: Implement governance frameworks and appoint data stewards to oversee master data management. Allow data steward to act with the Ma'at principle of Justice having the final say for use of data. Leadership should establish an open communication path with stewards to meet the needs of stewards and stakeholders.

Data Integration

Key Activities:

- **Integration Strategy**: Developing strategies for integrating data from various sources to achieve a unified view. Think about the enterprise and the needs of individual programs.
- **Consistency and Accuracy**: Ensuring that integrated data is consistent and accurate across systems. When software tools for visualizations are introduced, there should be little manipulation of raw data except for data scientists who are researching ways to streamline

operations and look for root cause analysis. The less touches on raw data, the better the output consistency.

- **Monitoring and Validation**: Continuously monitoring and validating data integration processes to address issues and improve accuracy.

Analytical Insights: Data integration strategies were pivotal in harmonizing data from various sources, providing a unified view of information. Lucy, Carl, and Janet's approach highlighted the need for consistency and accuracy in integrated data, supported by ongoing monitoring and validation to address any issues and enhance data integration processes.

Recommendations for Organizations:

- **Develop Integration Strategies**: Create comprehensive strategies for integrating data and communication mechanisms to link and discuss data lineage ensuring consultation with privacy controls.
- **Ensure Accuracy**: Focus on maintaining consistency and accuracy in integrated data. Create a culture that can go to data stewards and ask why information looks inaccurate and work on identifying if a policy change affected the expected results.
- **Monitor and Validate**: Continuously monitor and validate data integration processes to ensure data quality. Create cross functional teams to validate data. If you have coded data to produce a report the best

critique is an employee from another business process who is not intimate with the data. You want true validation not a check the box validation.

Cloud Integration

Key Activities:

- **Cloud Strategy and Planning**: Developing a strategy for cloud integration that aligns with organizational goals and the budget actuals, not projection. Involve Risk Management, Strategic Planning team, Technology office and the Program before making any purchases.
- **Data Migration and Management**: Managing data migration to the cloud and ensuring effective data management practices. But most importantly it requires a roadmap and strategy that won't sit on the shelf but one that will be used to check progress and milestones despite who will be in leadership. This plan should be so solid that if leadership changes, the strategy does not.
- **Security and Compliance**: Implementing security measures and ensuring compliance with data protection regulations in the cloud environment.

Analytical Insights: Cloud integration was approached with a strategic mindset, focusing on aligning cloud initiatives with organizational objectives. Effective data migration and management practices were crucial for a smooth transition to the cloud. Security and compliance measures ensured that data remained protected and met regulatory requirements.

Recommendations for Organizations:

- **Strategize for Cloud**: Develop a cloud integration strategy that aligns with organizational goals and objectives.
- **Manage Migration**: Oversee data migration and management to ensure a smooth transition to the cloud.
- **Secure and Comply**: Implement robust security measures and ensure compliance with data protection regulations in the cloud environment.

Now let's wrap up!

The practices of Lucy, Carl, and Janet in Harmony Village offer a valuable framework for achieving balance in data management. By applying principles of balance and justice to reference data management, master data consolidation, data integration, and cloud integration, organizations can enhance data consistency, accuracy, and security. The insights from their practices highlight the importance of defining, governing, consolidating, and integrating data effectively while maintaining a strong focus on quality and compliance. By following these recommendations, organizations can find their own promising practices and achieve a balanced and effective approach to data management.

Reflective Conversation

What does balance look like are your organization? Does it need work? Do your workers feel empowered with the tools they need to perform their work?

Principle of Balance: The principle of Balance in MDM emphasizes the harmonization and consistent management of data across all systems. It seeks to ensure that data integration and standardization efforts lead to a unified and coherent data environment, supporting effective decision-making and operational efficiency.

MDM Principle	Why It Fits the Principle of Balance	Actions for MDM Foundation
Reference Data Management	Standardizes key data to ensure consistency across systems.	Develop and maintain consistent reference datasets accessible to all systems.
Master Data Consolidation	Integrates data from multiple sources to create a single source of truth.	Utilize technologies to merge and reconcile data, eliminating duplicates.
Data Integration	Ensures seamless data flow between systems for operational harmony.	Implement integration tools and practices that support efficient data exchange.
Cloud Data Management	Manages data across cloud platforms to maintain system coherence.	Formulate strategies that ensure data consistency across on-premises and cloud platforms.

1. How can we ensure comprehensive and effective reference data management across our systems to standardize key

data and ensure consistency? KPI recommendations to test accountability:

+ Percentage of Reference Data Standardized: Measures the extent to which key data elements are standardized across systems.
+ Number of Systems with Access to Consistent Reference Data: Tracks how many systems are utilizing the standardized reference datasets.
+ Frequency of Updates to Reference Data Management

2. Are our master data consolidation efforts successfully creating a single source of truth by integrating data from multiple sources? KPI recommendations to test accountability:

+ Reduction in Data Redundancy and Duplication Rates: Evaluates the success of consolidation efforts in minimizing duplicate and redundant data.
+ Percentage of Data Sources Successfully Consolidated: Measures the proportion of data sources integrated into the master data repository.
+ Improvement in Reporting Accuracy Post-Consolidation: Assesses how well consolidation enhances the accuracy and reliability of reporting.

3. How effectively are we managing data integration to ensure seamless data flow and operational harmony across systems? KPI recommendations to test accountability:

+ Percentage of Data Integration Projects Completed on Time and Within Budget: Indicates the efficiency of completing integration projects according to schedule and financial constraints.

+ Number of Integration Issues Identified and Resolved: Tracks the effectiveness of addressing and resolving issues encountered during data integration.
+ Improvement in Data Flow Efficiency: Measures the enhancement in the speed and accuracy of data exchange between systems.

4. Are our cloud data management strategies effectively maintaining data consistency and coherence across on-premises and cloud platforms? KPI recommendations to test accountability:

+ Percentage of Data Consistent Across Cloud and On-Premises Platforms: Evaluates how well data consistency is maintained between cloud and on-premises systems.
+ Number of Cloud Data Management Issues Resolved: Tracks the effectiveness in addressing and solving problems related to cloud data management.
+ Frequency of Cloud Data Management Strategy Reviews: Assesses how often cloud data management strategies are audited for policy changes, how often are they reviewed and updated to ensure continued effectiveness?

Implementation Tip

Put in the work! Organizations must want and be committed to achieving balance. It has become a way of life, a culture that is influenced by leadership.

To achieve balance in data management, organizations should focus on developing and implementing practices that ensure consistency, accuracy, and efficiency. By asking the right questions and tracking relevant KPIs, organizations can assess and improve their reference data management, master data consolidation, data integration, and cloud data management strategies. This approach not only supports operational harmony but also helps maintain a balanced and effective data environment.

Chapter 6: Closeout

Synergize

Σ

In the heart of Harmony Village, a realm where the rivers of wisdom flow seamlessly and the principles of Ma'at—truth, balance, and justice—are lived out daily, the transformative power of exemplary data practices had long been a beacon for others. Lucy's journey began in the sterile environment of a doctor's office, where she first honed her skills in managing and safeguarding data. Her training, however, evolved and expanded, influencing not just her immediate environment but extending its impact to Harmony Village. As Harmony Village's methods proved to be a model of data management excellence, the village's practices began to ripple outwards, inspiring a wave of change beyond its borders.

Inspired by Harmony Village's success, Lucy, Carl, and Janet worked together to share these practices with a broader audience. Guided by the collective wisdom of the village elders and their own insights, they documented and demonstrated how the village's data management principles could be adapted and implemented in other organizations. They showcased how effective reference data management, master data consolidation, data integration, and cloud management could create harmonious data environments across diverse settings.

Summary

The journey of Harmony Village, shaped by Lucy's early experiences, Carl's technical expertise, and Janet's commitment to data protection, serves as a powerful testament to the principles of Ma'at. Their work exemplifies how embedding truth, balance, and justice into data management practices can lead to profound organizational success.

Truth: Harmony Village's practices demonstrate the importance of transparency and accuracy in data management. By implementing comprehensive documentation and effective auditing, they ensure that data remains accurate and reliable, which is essential for making informed decisions and maintaining trust.

Balance: The integration of data from various sources and the development of cohesive frameworks reflects the principle of balance. Harmony Village's approach ensures that data is harmonized across systems, avoiding over-reliance on any single source and providing a comprehensive view that supports operational harmony.

Justice: The ethical use of data, respect for privacy, and equitable decision-making are central to the village's practices. By adhering to strong data governance policies and promoting fairness, they uphold the principle of justice, ensuring that data management practices do not harm individuals or communities but rather drive positive outcomes.

Lessons Learned: From their experiences, organizations can learn the value of applying Ma'at's principles in their data practices. The successful replication of Harmony Village's methods demonstrates that by focusing on truth, balance, and justice, organizations can create data environments that are not only efficient and effective but also ethical and supportive of both employees and the communities they serve.

Through continuous monitoring, improvement, and the commitment to these core principles, **organizations can achieve a harmonious balance in their data management practices,** mirroring the success and wisdom of Harmony Village.

As we conclude this book, remember that the journey of integrating the principles of Ma'at into data management is just beginning. Throughout these pages, we've explored various ways to communicate principles and provided recommendations for embodying Ma'at in modern practices. You've seen different delivery styles and approaches, each chosen intentionally to offer a broad perspective.

What I hoped you learned:

Use of plain language and digestible content. Always ensure that the information is presented in way to make the concepts accessible and easily applicable. By breaking down complex ideas into digestible components, you enable employees to understand and implement the principles of Ma'at effectively.

Most importantly, I hope that the different styles prompted thought. Organizations must prioritize outcomes over rigid processes. Embrace innovation, test what works, and remain observant to identify and address what doesn't. This approach will enable you to adapt and refine your data practices to better serve your goals and values.

In the spirit of continuous improvement and sharing, the next section provides resources to help you create and document a framework for effective data management. These tools and references will guide you as you implement the principles of Ma'at and strive for balance, justice, and truth in your organization.

Happy harmonizing! Now let's go serve the greater good and drive positive change.

Resources:

MDM Roadmap

Phase 1: Establishing the Foundation (Months 1-3)

Focus: Truth and Justice

1. **Data Discovery and Documentation**
    - **Action:** Conduct data discovery sessions to identify all data sources.
    - **Output:** A comprehensive inventory of data sources and data types.
    - **Goal:** Ensure all data is accounted for and its purpose is understood.
2. **Develop Data Governance Framework**
    - **Action:** Establish data governance policies, including data ownership, access rights, and compliance requirements.
    - **Output:** A documented governance framework.
    - **Goal:** Set the rules and responsibilities for data management.
3. **Define Data Quality Rules**
    - **Action:** Establish data quality criteria and validation processes.
    - **Output:** Data quality guidelines and automated validation scripts.
    - **Goal:** Ensure the accuracy and consistency of data from the beginning.
4. **Set Up Metadata Management**

- **Action:** Define metadata standards and implement a metadata management tool.
- **Output:** A centralized metadata repository.
- **Goal:** Provide context and transparency for all data within the organization.

Phase 2: Building Accountability and Security (Months 4-6)

Focus: Justice

5. **Implement Data Lineage and Auditing**
    - **Action:** Document the flow of data from ingestion to usage, including transformation steps.
    - **Output:** A data lineage and audit trail.
    - **Goal:** Ensure transparency and accountability for data handling.
6. **Appoint Data Stewards**
    - **Action:** Identify and assign data stewards to oversee data quality and governance.
    - **Output:** A roster of data stewards with clearly defined roles.
    - **Goal:** Promote ethical and responsible data management.
7. **Establish Data Security and Privacy Protocols**
    - **Action:** Implement data security measures, including encryption, access controls, and compliance with privacy regulations.
    - **Output:** A comprehensive data security policy.

- **Goal:** Protect sensitive data and ensure legal compliance.

Phase 3: Achieving Balance and Consistency (Months 7-9)

Focus: Balance

8. **Standardize Reference Data**
    - **Action:** Develop a standardized set of reference data for use across the organization.
    - **Output:** A centralized reference data repository.
    - **Goal:** Ensure consistency and avoid discrepancies in key data elements.
9. **Master Data Consolidation Planning**
    - **Action:** Create a strategy for consolidating master data, focusing on creating a single authoritative source for key business entities.
    - **Output:** A master data consolidation plan.
    - **Goal:** Achieve a unified view of critical business data.
10. **Plan for Cloud Data Integration**
    - **Action:** Develop and implement a data integration strategy that supports seamless data flow between on-premises and cloud systems.
    - **Output:** A cloud integration roadmap.
    - **Goal:** Ensure data accessibility and consistency across all platforms.

Phase 4: Ongoing Management and Evolution (Months 10-12 and Beyond)

Focus: Continuous Improvement and Scalability

11. **Implement Data Lifecycle Management**
    - **Action:** Establish policies for data retention, archiving, and secure disposal.
    - **Output:** A data lifecycle management policy.
    - **Goal:** Manage data effectively from creation to disposal, ensuring compliance and efficiency.
12. **Develop Cloud Data Management Plan**
    - **Action:** Implement a plan for managing data in cloud environments, including strategies for backup, disaster recovery, and compliance.
    - **Output:** A comprehensive cloud data management strategy.
    - **Goal:** Support scalability and flexibility as your business grows and transitions fully to cloud technology.

Master Data Management Policy Framework: Aligning with the Principles of Ma'at

This policy framework outlines the foundational principles and guidelines for implementing and maintaining Master Data Management (MDM) in alignment with the principles of Ma'at: Truth, Justice, and Balance. This framework is designed to ensure that data is managed with integrity, accountability, and harmony, supporting the growth and success of the organization.

1. Truth: Ensuring Data Accuracy and Transparency

Objective:
To establish a data environment where accuracy, consistency, and transparency are prioritized, ensuring that all data is trustworthy and reliable.

Key Policies:

- **Data Discovery and Documentation Policy:**
  - **Purpose:** To identify and document all data sources and their usage within the organization.
  - **Guidelines:** All departments must conduct thorough data discovery sessions and maintain up-to-date documentation of data sources, types, and purposes.
- **Data Governance Policy:**
  - **Purpose:** To establish a clear governance structure for data management across the organization.
  - **Guidelines:** Implement governance policies that define data ownership, access rights, and compliance requirements. Regularly review and update these policies to align with evolving business needs and regulations.
- **Data Quality Policy:**
  - **Purpose:** To ensure that all data meets defined quality standards for accuracy, completeness, and consistency.
  - **Guidelines:** Develop and enforce data quality rules. Implement automated validation

processes to ensure that only high-quality data is ingested and maintained.

- **Metadata Management Policy:**
    - **Purpose:** To provide context and understanding of the data by managing metadata effectively.
    - **Guidelines:** Establish metadata standards and ensure that all data attributes, definitions, and sources are documented in a centralized metadata repository.

2. Justice: Ensuring Accountability and Ethical Data Management

Objective:

To promote fairness, accountability, and ethical data practices throughout the organization, ensuring that data is managed responsibly and securely.

Key Policies:

- **Data Lineage and Auditing Policy:**
    - **Purpose:** To maintain transparency in data handling by tracking the flow and transformation of data across the organization.
    - **Guidelines:** Document and audit data lineage from ingestion to storage. Ensure that all data transformations are traceable and verifiable.
- **Data Stewardship Policy:**

- **Purpose:** To assign accountability for data quality and governance within the organization.
- **Guidelines:** Appoint data stewards with clearly defined roles and responsibilities. Data stewards are responsible for overseeing data quality, compliance, and governance activities.

- **Data Security and Privacy Policy:**
  - **Purpose:** To protect sensitive data and ensure compliance with legal and ethical standards.
  - **Guidelines:** Implement strong data security measures, including encryption, access controls, and compliance with data privacy regulations (e.g., GDPR, CCPA). Regularly review and update security protocols to address emerging threats.

3. Balance: Ensuring Consistency and Harmonious Data Integration

Objective:
To create a balanced and integrated data environment that supports the seamless flow and consistency of data across all systems and platforms.

Key Policies:

- **Reference Data Management Policy:**
  - **Purpose:** To standardize and maintain consistency in reference data across all systems.

- **Guidelines:** Develop and maintain a centralized reference data repository. Ensure that all systems use the standardized reference data to avoid discrepancies.
- **Master Data Consolidation Policy:**
  - **Purpose:** To create a unified view of critical business entities by consolidating master data.
  - **Guidelines:** Implement a strategy for consolidating master data, focusing on creating and maintaining a single, authoritative source for key data entities.
- **Data Integration Policy:**
  - **Purpose:** To facilitate the seamless flow of data between on-premises and cloud systems.
  - **Guidelines:** Develop a robust data integration strategy that supports real-time and batch processing. Ensure that data is consistent and accessible across all platforms.
- **Data Lifecycle Management Policy:**
  - **Purpose:** To manage data effectively throughout its lifecycle, from creation to disposal.
  - **Guidelines:** Establish data retention, archival, and disposal policies that align with business needs and legal requirements. Ensure that data is securely archived or disposed of when no longer needed.
- **Cloud Data Management Policy:**
  - **Purpose:** To ensure effective management of data in cloud environments.

- **Guidelines:** Implement a cloud data management plan that includes strategies for data backup, disaster recovery, and compliance with cloud-specific regulations. Ensure that cloud data management practices align with overall MDM strategies.

Implementation:

All departments must adhere to the MDM policies outlined in this framework. Training and awareness programs should be conducted regularly to ensure that all employees understand and comply with these policies.

Review and Continuous Improvement:

The MDM policies must be reviewed annually or as required by changes in business needs, technology, or regulations. Continuous improvement practices should be implemented to refine and enhance MDM processes, ensuring alignment with the principles of Ma'at.

Governance and Compliance:

A dedicated MDM governance committee should be established to oversee the implementation of this policy framework. The committee is responsible for monitoring compliance, addressing issues, and making recommendations for policy updates.

This policy framework ensures that your organization's MDM practices are aligned with the principles of Truth, Justice, and

Balance, fostering a culture of integrity, accountability, and harmony in data management.

Project Management Plan for Developing a New System Aligned with Ma'at Principles

Objective:

To develop a new data management system that embodies the principles of Ma'at: Truth, Justice, and Balance, ensuring that the system is built on a foundation of accurate, ethical, and consistent data management practices. The project emphasizes the importance of gathering the right requirements to align the system with these principles and meet business needs effectively.

1. Project Overview

Project Name:
Ma'at-Aligned Data Management System (MDS)

Project Duration:
12 months

Project Team:

- **Project Sponsor:** Chief Data Officer (CDO)
- **Project Manager:** Senior IT Project Manager
- **Business Analyst:** Lead Business Analyst
- **Technical Lead:** Senior Systems Architect
- **Data Governance Lead:** Data Steward

- **Development Team:** Software Engineers, Database Administrators, UI/UX Designers
- **Quality Assurance:** QA Lead and Testers
- **End Users:** Representatives from business units

2. Project Goals

- **Truth:** Ensure the system provides accurate, reliable, and transparent data by clearly defining and understanding business requirements.
- **Justice:** Develop the system with ethical considerations in mind, ensuring that data is managed responsibly and securely.
- **Balance:** Create a harmonious data environment where consistency and integration across platforms are prioritized.

3. Project Phases

Phase 1: Initiation and Planning (Months 1-2)

Key Activities:

Stakeholder Identification and Engagement: Identify all stakeholders and ensure they are engaged in the planning process.

Requirements Gathering:

- **Purpose:** Gather detailed and accurate requirements from all relevant stakeholders.

- **Why it Matters:** Gathering the right requirements is crucial because it ensures that the system will meet the actual needs of the business and align with the principles of Ma'at. Poorly defined requirements can lead to project delays, increased costs, and a system that fails to deliver value.
- **Activities:**
    - Conduct workshops, interviews, and surveys with stakeholders.
    - Develop use cases and user stories to capture detailed requirements.
    - Prioritize requirements based on business value and alignment with Ma'at principles.
- **Output:** A comprehensive Requirements Specification Document.

Deliverables:

- Project Charter
- Requirements Specification Document
- Project Plan (including timelines, milestones, and resource allocation)

Phase 2: System Design (Months 3-4)

Key Activities:

System Architecture Design:

- **Purpose:** Design the system architecture that ensures data accuracy, security, and consistency.

- **Why it Matters:** A well-designed architecture is the backbone of the system, ensuring that it can support accurate data management (Truth), uphold security and compliance (Justice), and provide a consistent user experience (Balance).
- **Activities:**
  - Define data models, workflows, and integration points.
  - Design security protocols and data governance structures.
  - Create a blueprint for user interface (UI) and user experience (UX).

Deliverables:

- System Architecture Document
- Data Governance Framework
- UI/UX Design Prototypes

Phase 3: Development (Months 5-8)

Key Activities:

System Development:

- **Purpose:** Develop the system according to the design specifications, ensuring that it meets the requirements and adheres to Ma'at principles.
- **Why it Matters:** The development phase translates the requirements and design into a functional system. Adhering to the principles of Truth, Justice, and

Balance ensures that the system will operate with integrity and provide real value.

- **Activities:**
    - Develop core system functionalities, including data ingestion, processing, and storage.
    - Implement data validation and quality control mechanisms.
    - Integrate security features and compliance checks.
    - Develop reporting and analytics modules to provide transparency.

Deliverables:

- Functional System Modules
- Data Validation and Quality Control Scripts
- Security and Compliance Features
- Reporting and Analytics Dashboard

Phase 4: Testing and Quality Assurance (Months 9-10)

Key Activities:

Testing and QA:

- **Purpose:** Ensure that the system functions as intended, is free of defects, and aligns with Ma'at principles.
- **Why it Matters:** Rigorous testing and quality assurance are critical to delivering a system that users can trust (Truth), that protects sensitive data (Justice), and that provides a consistent experience (Balance).
- **Activities:**

- Conduct unit, integration, and system testing.
- Perform user acceptance testing (UAT) with key stakeholders.
- Validate data accuracy, security, and system performance.

Deliverables:

- Test Plans and Test Cases
- Testing Reports and Bug Fixes
- User Acceptance Test (UAT) Sign-Off

Phase 5: Deployment and Training (Months 11-12)

Key Activities:

System Deployment:

- **Purpose:** Deploy the system to the production environment and ensure it is fully operational.
- **Why it Matters:** Deployment is the final step in making the system available to users. A successful deployment ensures that the system meets its objectives and aligns with Ma'at principles in real-world use.
- **Activities:**
    - Deploy the system to the production environment.
    - Migrate data from legacy systems, ensuring data integrity.
    - Monitor system performance post-deployment.

- **User Training:**
    - **Purpose:** Equip users with the knowledge and skills to use the system effectively.
    - **Why it Matters:** Proper training ensures that users can interact with the system efficiently and in a way that upholds the principles of Ma'at.
    - **Activities:**
        - Conduct training sessions for end users, focusing on key functionalities.
        - Provide user manuals and support materials.
        - Offer ongoing support and helpdesk services.

Deliverables:

- Deployed System
- Data Migration Reports
- Training Materials and User Manuals
- Support and Maintenance Plan

4. Importance of Gathering the Right Requirements

Gathering the right requirements is the foundation of any successful system development project. In the context of this MDM system, gathering accurate and detailed requirements is crucial for the following reasons:

13. Alignment with Ma'at Principles:

- **Truth:** Accurate requirements ensure that the system provides reliable data, reflecting the true state of the business.
- **Justice:** Clear requirements help define ethical data practices, ensuring that the system upholds privacy, security, and compliance standards.
- **Balance:** Well-defined requirements ensure that the system integrates seamlessly with existing processes and provides a consistent user experience.

14. **Risk Mitigation:**
    - Misunderstood or poorly defined requirements are a common cause of project failure. By gathering the right requirements upfront, the project team can mitigate risks such as scope creep, delays, and cost overruns.
15. **User Satisfaction:**
    - The system must meet the needs of its users to be successful. By involving users in the requirements-gathering process, the project team ensures that the system will be user-friendly and meet business needs.
16. **Cost Efficiency:**
    - Clear requirements help avoid rework and reduce the likelihood of costly changes later in the project. This contributes to a more efficient use of resources and a higher return on investment.
17. **Scalability and Futureproofing:**
    - Gathering comprehensive requirements ensures that the system is designed with future growth in mind. This allows for scalability and adaptability as business needs evolve.

5. Conclusion

This project management plan outlines a structured approach to developing a new data management system that is aligned with the principles of Ma'at. By emphasizing the importance of gathering the right requirements, the plan ensures that the system will meet business needs, operate with integrity, and provide long-term value. Through careful planning, design, development, and deployment, the project team will deliver a system that embodies the principles of Truth, Justice, and Balance, fostering a harmonious data environment for the organization.

Project Charter Template

1. Project Overview

Project Name:
[Insert Project Name]

Project Sponsor:
[Insert Name and Title of Project Sponsor]

Project Manager:
[Insert Name and Title of Project Manager]

Project Start Date:
[Insert Start Date]

Expected End Date:
[Insert End Date]

2. Project Purpose and Justification

Business Need/Problem Statement:

- Clearly define the problem or opportunity that the project aims to address.
- Explain the importance of the project in relation to the organization's strategic goals.

Project Objectives:

- List the specific, measurable objectives the project aims to achieve.
- Ensure each objective aligns with the overarching business need.

3. Project Scope

Scope Description:

- Provide a detailed description of what is included within the scope of the project.
- Outline the key deliverables and the high-level approach to delivering them.

Out of Scope:

- Clearly define what is not included in the scope of the project to manage expectations.

4. High-Level Requirements

- Outline the key business and technical requirements that the project must fulfill.
- Specify any compliance, security, or regulatory requirements.

5. Key Stakeholders

Stakeholder Identification:

- List the key stakeholders involved in or affected by the project.

Stakeholder Name	Role	Responsibilities	Contact Information
[Name]	[Role/Title]	[List of Responsibilities]	[Email/Phone]
[Name]	[Role/Title]	[List of Responsibilities]	[Email/Phone]

6. Project Milestones

Milestone	Target Date	Description
[Milestone Name]	[Date]	[Brief description of milestone]
[Milestone Name]	[Date]	[Brief description of milestone]

7. Project Budget

Estimated Budget:

- Provide an overview of the project budget.

Resource Allocation:

- List the resources required, including personnel, equipment, software, and any other expenses.

8. Project Risks and Mitigation Strategies

Risk	**Impact**	**Likelihood**	**Mitigation Strategy**
[Risk Description]	[High/Medium/ Low]	[High/Medium/ Low]	[Strategy to mitigate the risk]
[Risk Description]	[High/Medium/ Low]	[High/Medium/ Low]	[Strategy to mitigate the risk]

9. Project Constraints

- Identify any limitations, such as time, budget, resource availability, or technology constraints that may impact the project.

10. Project Assumptions

- Document any assumptions made during the planning process that impact the project's scope, schedule, or resources.

11. Project Approval

Approval Signatures:

- The following stakeholders approve the project charter and agree to support the project through to its completion.

Name	Role	Signature	Date
[Name]	[Role/Title]	[Signature]	[Date]
[Name]	[Role/Title]	[Signature]	[Date]

12. Communication Plan

Communication Strategies:

- Define how communication will be managed throughout the project lifecycle.
- Identify key communication methods, frequency, and audiences.

13. Change Management Plan

Change Control Process:

- Outline the process for managing changes to the project scope, schedule, or budget.
- Include who has the authority to approve changes and how they will be documented.

This template serves as a foundational document for project initiation and planning, providing clear guidance and alignment among stakeholders. Customize it according to the specific needs of your project and organization.

Communications Plan Framework

Purpose and Objectives

- **Purpose:** Define the overall goal of the communication plan and explain how it supports the project's objectives.
- **Objectives:** Outline specific aims such as keeping stakeholders informed, ensuring transparency, fostering collaboration, and managing expectations.

2. Key Stakeholders

- **Stakeholder Identification:** List all stakeholders involved or affected by the project, including both internal (e.g., team members, management) and external (e.g., customers, vendors).
- **Stakeholder Communication Needs:** Detail the communication needs and preferences for each stakeholder group, specifying different requirements for senior management versus team members.

3. Communication Methods

- **Channels of Communication:** Identify the communication channels to be used (e.g., email, meetings, newsletters) and specify when and how each will be utilized based on stakeholder needs.
- **Communication Tools:** List tools supporting communication (e.g., project management software, video conferencing) and explain their use.

4. Communication Schedule

- **Frequency of Communication:** Define the frequency of communications for various types (e.g., daily stand-ups, weekly updates) to align with stakeholder needs.
- **Communication Calendar:** Develop a calendar detailing key communication activities, including milestones, meetings, and reports with dates and times.

5. Message Content and Delivery

- **Key Messages:** Identify and tailor key messages for different audiences while maintaining overall consistency.
- **Message Format:** Define the format of messages (e.g., presentations, reports, emails) to ensure appropriateness for content and audience.

6. Roles and Responsibilities

- **Communication Roles:** Assign roles and responsibilities for communication tasks (e.g., project manager, communication coordinator) and specify who is responsible for creating, reviewing, approving, and delivering communications.
- **Approval Process:** Outline the review and approval process for communications, ensuring all key stakeholders can review critical messages.

7. Feedback Mechanisms

- **Feedback Collection:** Define methods for collecting feedback from stakeholders (e.g., surveys, feedback forms) and ensure opportunities for input.
- **Feedback Analysis:** Develop a process for analyzing feedback and incorporating it into future communications to enhance effectiveness.

8. Evaluation and Metrics

- **Communication Effectiveness:** Define metrics for assessing the effectiveness of the communication plan (e.g., stakeholder satisfaction, message comprehension) and use these to evaluate success and identify improvement areas.
- **Continuous Improvement:** Establish a process for regularly reviewing and updating the communication plan based on feedback and evaluation results.

9. Risk Management

- **Communication Risks:** Identify potential risks impacting communication (e.g., miscommunication, information overload) and develop mitigation strategies.
- **Contingency Plans:** Outline contingency plans for addressing communication issues that arise during the project, ensuring quick resolution.

10. Documentation and Archiving

- **Communication Records:** Establish processes for documenting and archiving all project-related communications, ensuring secure storage and easy access.

- **Knowledge Management:** Develop strategies for sharing and retaining knowledge gained from the project, documenting key lessons learned for relevant stakeholders.

This framework provides a structured approach to crafting a communication plan that ensures clarity, consistency, and effectiveness throughout the project lifecycle. Adapt it to fit the specific needs of your project and organization.

References:

Amen, Ra Un Nefer. *Ma'at: The 11 Laws of God.* Inner Traditions, 2003.

Knight, Michelle. "What Is Data Architecture? Components and Uses." *DATAVERSITY*, 2 Nov. 2023, www.dataversity.net/what-is-data-architecture/.

National Institute of Standards and Technology. *NIST Glossary of Statistical Terms.* National Institute of Standards and Technology, n.d. Web. 10 Aug. 2024.

"What Is Data Modeling?" *IBM*, 29 Sept. 2021, www.ibm.com/topics/data-modeling.

UKMT Morality and Ethics Collective. *Ma'at: Guiding Principles of Moral Living.* UKMT Publishing, 2014.

Made in United States
Orlando, FL
04 November 2026

90201130R00065